SILVER ISLET

Its original name was Skull Rock.

Barely awash in the frigid waters of Gitchee-Gumee, a few kilometres from Nanabijou's southern flank, it was stumbled upon one summer's day in 1868 by surveyors of the Montreal Mining Company, who unearthed silver-bearing ore and a rich underwater vein. But the feasibility of deep-shaft mining on such a remote and seagirt atoll, and in such hostile surroundings, was, to say the least, doubtful. Not until the fall of 1870 did a courageous group of entrepreneurs, calling themselves the Silver Islet Mining Company of Lake Superior and led by Major A. H. Sibley, president, and William B. Frue, superintendent, acquire title to Skull Rock and prepare to attempt the impossible.

That they succeeded is by now of course a matter of history. At cessation of operations in 1884, Skull Rock had been enlarged tenfold by means of submerged cribbings and timbered sea walls. The inclined shaft had been sunk below 1200 feet, the vertical shaft to almost 700. Over three and a quarter million dollars (in today's currency fifty times that amount) worth of silver ore had been extracted.

During the glory years, the mainland community flourished and became a staging point between eastern and western Canada, with socio-economic links to Toronto, Detroit, New York, Duluth and Winnipeg. Mansions, wharves, churches, a general store, a hotel, a jail and a hundred homes were built; and whereas the head frames, sorting sheds and dormitories on the Islet have long since disappeared, destroyed by a century of Lake Superior storms, many of the original mainland structures still stand, and appear in these photographs.

- from the Preface,
CONFESSIONS OF A CORNISH MINER
James Strathbogey

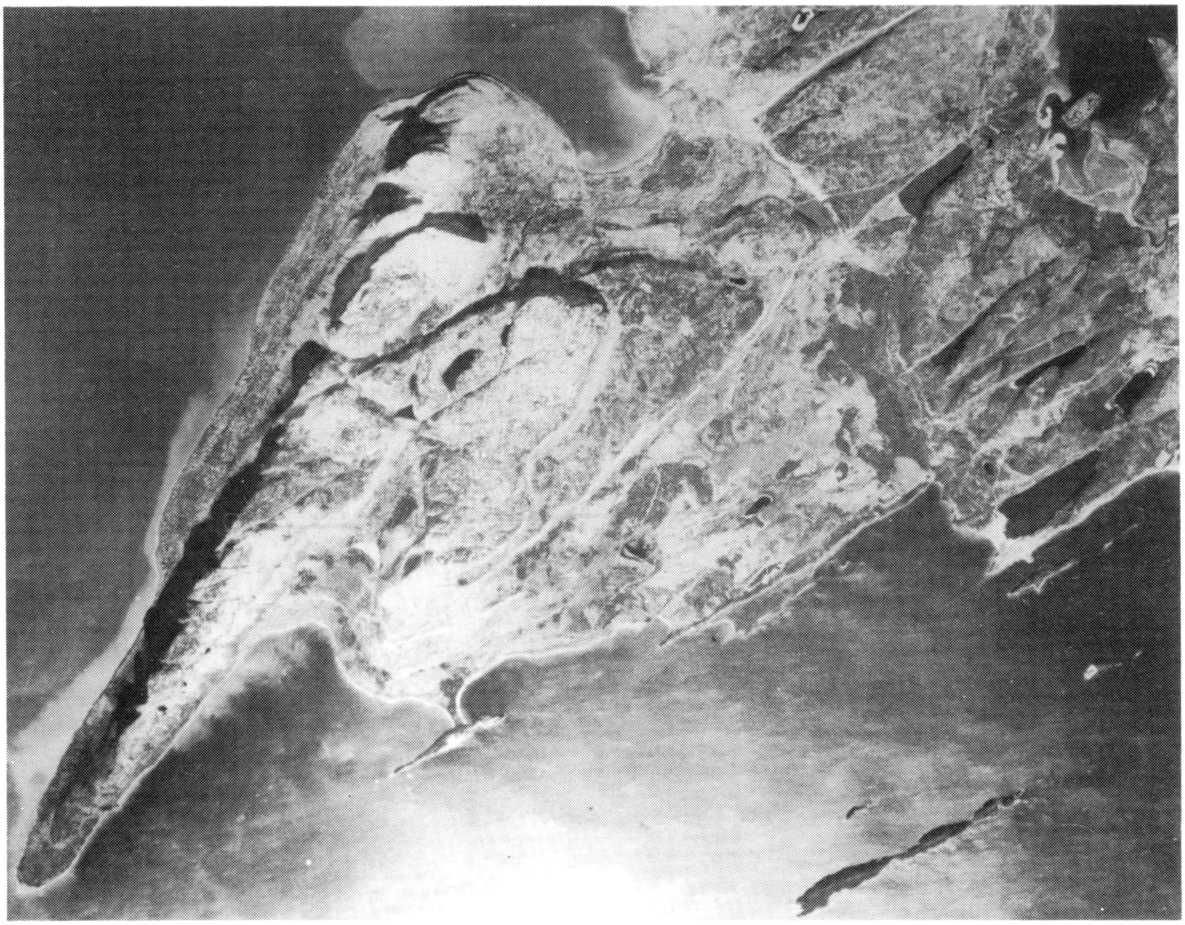

Aerial view of Sibley Peninsula & Sleeping Giant Provincial Park.
Silver Islet & the Woods Location at extreme right.

"Included is an oblong tract, five miles in length coastwise, and two miles in width, containing 64 hundred acres, designated and known as the Woods Location. It is situated five miles easterly of Thunder Cape and twenty miles northerly from the eastern end of Isle Royale, and is traversed in a southwesterly direction by a silver-bearing lode of demonstrated richness and promise. September last, when possession was obtained, exploration of this vein was commenced where it was exposed on a small island of diorite, now known as Silver Islet. The vigorous prosecution of the enterprise which was urgently invited by these explorations had forbidding aspects: the locality was remote, wild and desolate, the season far advanced, the time of storms at hand, and winter approaching. It was determined, however, that the effort should be made, and great as was the hazard, it enhances the credit due to the managers, Messrs. Sibley and Frue, who have materially contributed to bringing the enterprise to its now realized success."

- Edward Learned, Treasurer
Silver Islet Mining Company
6th February, 1871

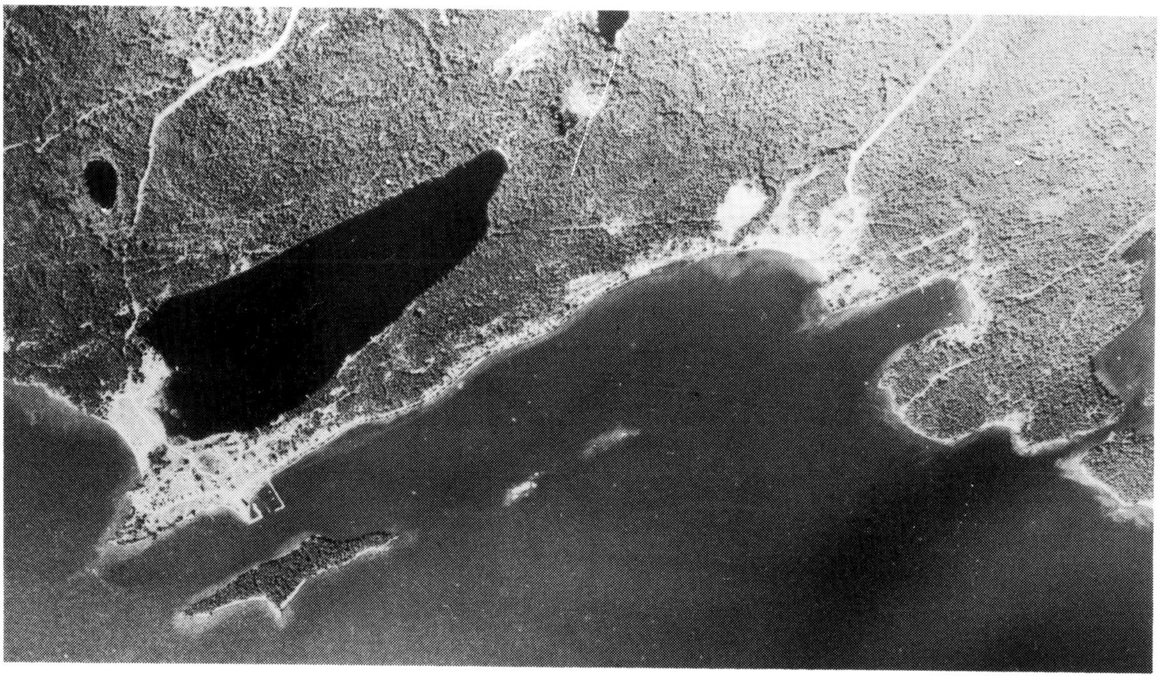

Aerial view of the mainland community at Silver Islet.
Surprise Lake left centre. Avenue, store & dock foreground.
Camp Bay & entrance to Fork Bay extreme right.
Burnt Island & Powder Rock lower left.

"This undertaking involved the procuring of the necessary force of operatives and their sustenance, the construction of dwellings, shops and storehouses, means of transit to and from the mine, all utensils and appliances for mining operations, and protection of the mine from storms of the lake, from fields of floating ice without, and from leakage within, together with a multitude of articles of real or possible necessity. The force at the mine presently consists of 70 men; although as several have their families with them, the colony is somewhat larger. The practical miners have been selected from the copper regions on the south shore, with reference to their skill and worthiness. The entire force is under the supervision of Mr. Wm. B. Frue, a mining captain of conceded capacity and integrity."

- Edward Learned, Treasurer
Silver Islet Mining Company
6th February, 1871

"There have been received at Balbach's Smelting Works in Newark, N.J. 287 barrels of silver ore from Silver Islet, being the product of the mine during October and November, 1870. Estimating the ore at the same rate as that already treated, the yield will be $101,006 currency."

- Charles Trowbridge, Secretary
Silver Islet Mining Company
10th February, 1871

The Store. Note water barrels on roof.

"About December 1st, 1870, the shaft was reported to be eighteen feet deep. The depth to which the vein has been proved gives assurance that within the limit of the coffer dam alone there is a mass of ore estimated to be worth $800,000, while beyond such limit it is expected that a much greater quantity can be reached.

"In addition to the buildings, breakwaters and tools before referred to, supplies of steel, iron, powder, fuse, steam engines, steam tugs, scows, timber, horses, feed, furniture, and general subsistence stores have been paid for, sufficient to carry forward operations until May next. Hence, it may be assumed that when navigation opens, the ore on hand at the mine will substantially free us from all charges, other than for labour, which may amount to thirty-five or forty thousand dollars."

- Edward Learned, Treasurer
Silver Islet Mining Company
6th February, 1871

"On the 1st of September, 1870, I became the purchaser of these lands for myself and others, for $225,000, which has been paid in full. Captain Frue, on a telegram from me, at once chartered a steamer and left Houghton, Michigan, with thirty men, mining tools and supplies, for Silver Islet, towing a large scow and raft of timber to commence the construction of a breakwater.

"From that time until the 10th of May 1871, Captain Frue was engaged in constructing works to protect Silver Islet, at enormous cost. A breakwater was at first constructed which enabled him to begin mining in about 30 days. In November, 1870, a heavy gale carried away a part of this structure, but it was repaired and mining began again. On the 8th of March, 1871, a heavy sea, bearing a large body of ice, carried away half the crib work and filled the mine with water, and the works were not restored until the 10th of May.

"On the island and breakwater there are now erected a large boarding house, blacksmith's shop, engine house, sorting shed and shaft house, and mining operations have been carried on uninterruptedly since the 10th of May last. The number of men employed on and about Silver Islet is 176 in summer and 130 in winter, apart from the force employed in smelting. On the mainland we have erected forty dwellings, a church and schoolhouse, a large store for the sale of commodities, and have also built a harbour with breakwater, basins and wharves. There being no lighthouse on the coast, we have kept at our own expense a large light on Silver Islet and two range lights on the main shore, continually alight from dark to daylight for the benefit of steam and sailing craft plying these waters. Ours is really the only Harbour of Refuge in Thunder Bay District, as the masters of the *Chicora, Manitoba, Cumberland, Algoma* and other vessels can testify."

- A. H. Sibley, President
Silver Islet Mining Company
20th February, 1872

Some of the structures that once stood on the Islet.
Shaft house at extreme left.

“The *Mineral Rock* arrived here Wednesday evening last and will undoubtedly start for Marquette tomorrow morning, taking from here some 50 barrels of ore. Today her Captain and I went out to the Islet and took soundings at the most desirable place of hauling alongside. At the southeast end of our dock, which is the most convenient place for loading, the water is sufficiently deep, but there was no chance of staying the stem of the boat. These facts ascertained, together with the uncertainty of the weather, the Captain finally decided to run over to Marquette and take on a load of iron there. The weather this fall has been unusually boisterous. Every day we have a blow from some quarter. Tonight it is blowing quite a little a gale from the south and the sea is making fast.

"The building of the large new boarding house on the Islet has blocked out an immense amount of work, to accomplish which calls upon me for 30 thousand feet of timber and ten thousand tons of rock. The furnishing of this amount of material is no small undertaking at this season of the year, and will compel the running of the scow well into December."

- Wm. B. Frue, Silver Islet
20th October, 1871

"As it has been blowing a perfect hurricane all afternoon, and snowing very hard, all communication with the Islet has been cut off. The steamer *Manitoba* has been lying all day at our dock, unable to proceed further, owing to the severity of the weather. Our scows have been moored inside the breakwater since Monday and the mine will undoubtedly lie idle tonight."

- Wm. B. Frue, Silver Islet
26th October, 1871

Some of the buildings on Silver Islet, after the closure of the mine. The end of an era. Shaft house at extreme right.

"At the Islet, the large boarding house is surrounded by crib work from all exposed quarters. Our southern breakwater is now completed, with an elevation of 17 feet above lake level, and seems to have sufficient strength to withstand the forces of the watery elements, which however roll over the top in considerable quantity."

- Wm. B. Frue, Silver Islet
13th December, 1871

Boats of a bygone era: steam tugs at the dock.
Note the bell on the old warehouse at right, above the Gracie.

"The *Silver Spray* has gone into winter quarters protected by a new breakwater 150 feet east of the old one.

"The health of the community is not as good as I could wish: several cases of a type of typhus fever have occurred, although not of a fatal nature. The cases take 4 weeks to develop and appear to be a general poisoning of the blood."

- Wm. B. Frue, Silver Islet
19th January, 1872

Steam tugs at the dock.

Silver Islet sunrise.

"When at Prince Arthur's Landing the day before yesterday, I saw Mr. Van Norman, the Magistrate, who informed me that there was no affidavit to the execution of the deed of Silver Islet. He is not sure whether it was attached to the one he returned to Major Sibley.

"We are having a very pleasant winter here. The weather is delightful. The broad sheet of water between us and Isle Royale is coated with ice 15 inches thick, which formed last week."

- Wm. B. Frue, Silver Islet
4th February, 1872

The Beach.

"I am sorry to state that it would put me to a good deal of inconvenience to have Trethewey leave Silver Islet at the present time.

"Owing to the large amount of rock we are obliged to quarry this winter, we have used considerable powder, and it is a question whether I can spare the Shuniah Mine enough to be of any use. I furnished them with 50 kegs in December, which I understand is now exhausted."

- Wm. B. Frue, Silver Islet
13th February, 1872

A Disgruntled Cornish Miner?

"From a dissatisfaction among the men, chiefly the Cornish Miners, I am satisfied that but few will remain here upon the opening of navigation. To secure replacements, I am of the opinion that notices should be put into newspapers in different mining localities of the country. This would avoid sending around to drum up men and pay their passage, which makes them worthless. You pay a man's fare to a place, and as a general rule that man is no good to himself or to you. He labours under the delusion that you cannot possibly get along without him."

- Wm. B. Frue, Silver Islet
19th February, 1872

Disgruntled Silver Islet Miners.

Happy-Go-Lucky.

The Jail.

The Hotel.

"Last week we took out 37 barrels of good ore, valued at $6000 to the ton. The mine was idle a portion of the time, owing to a strike. This strike compelled me to settle with eight of the miners, all of whom went to the Landing."

- Wm. B. Frue, Silver Islet
26th February, 1872

"Catholic Point," foreground.

"The mine at the bottom continues to look well. The product for February was 100 barrels, which I think will be equal in value to the ore shipped last spring - say $1700 per ton. In addition, two barrels were taken out which ought to yield over $6000 to the ton. These two barrels alone will cover the working expenses of the month, which will leave the 100 barrels for the stockholders."

- Wm. B. Frue, Silver Islet
1st March, 1872

Trowbridge lighthouse from Camp Bay.

James Booker's ghost?

"The product for the month will be about 122 barrels of ore. The is a perceptible increase over last month, considering the difference in the size of the barrels, which were from the new stock and measure 42 gallons.

"The ice in the lake remains solid, being fully 3 feet thick and covered with 2 feet of snow. I do not expect any chance of a boat reaching here before the 1st of May, and there is little chance of Thunder Bay being open before June."

- Wm. B. Frue, Silver Islet
29th March, 1872

The President's house, shortly before it burned to the ground.

"The carpenters are busily engaged in putting on an addition to your house, to the southeast of the main building. The ground floor is divided into a dining room, a sitting room and a bedroom. The upper storey will be divided into four bedrooms. I contemplate taking down the partition between the sitting room and your bedroom of last year, and making a large parlour, or whatever you choose to call it. The sitting room in the southeast wing is intended for your own private quarters. The entrance to the dining room will be from the hall door facing on Sibley Cove. I believe I will have the house in such shape that Mrs. Sibley will be delighted with it, for it will accommodate any reasonable number of visitors without the slightest interruption of her own domestic quarters."

- Wm. B. Frue, Silver Islet
4th April, 1872

"The Store has done a fair amount of business during the winter. Of course trade is chiefly confined to the employees. In the summer the Store will do a large business if a good selection of saleable goods is purchased, such as silks, plaids, prints, etc. A large profit might be devised by selling to visitors who will flock here by the thousands and who will run the risk of smuggling into the United States. But all of these goods should be selected with the greatest care and taste.

"Some of the purchases made last fall will remain on hand for some time to come. I allude chiefly to clothing, scarcely any of which was sufficiently large for an ordinary sized man. Nearly all of the mining suits remain on the shelf for this reason. The German socks purchased from Austria have not proved as saleable an article as I expected.

"A few small orders for articles most urgently needed at the Store have gone forward, such as butter, sugar, etc., all of which are entirely out and we are unable to purchase any more up in this country. We still have a large supply of pork, flour, and such staple articles on hand.

"I am pleased to state that the vein at the Islet looks more promising that it has at any time in the past four months. We have been sinking rapidly for the last 3 weeks and I feel quite pleased at the general appearance of things. The ice is still solid in the lake, the weather fine and warm for this season of the year."

- Wm. B. Frue, Silver Islet
4th April, 1872

Captain Cross - progenitor.

STORE.
ICE
PEPSI
Crush

"Should Mr. Ward throw his stock upon the market to dispose of same, he himself will be the only one who will lose, if my prediction for the coming summer should be anywhere near realized.

"Last night we were visited with a storm of thunder, lightning, hail, snow and rain, by way of variety. At present there are no indications of an opening of navigation before the 1st of May.

"Mr. Davidson has got his list of purchases ready and will leave here on the first boat for Toronto. Sales made at the Store to date amount to $18,000. A considerable quantity of stock now on hand consists of staple articles, so that the purchase of provisions for the summer will be quite light."

- Wm. B. Frue, Silver Islet
12th April, 1872

Captain Frue's house at left: "Burry Worry".

Vessels docked at Silver Islet, circa 1910. Note the Catholic church at left, just visible beyond the Simcoe's *stern.*

"The steamer *Manitoba* arrived here this morning, being the first boat of the season. I shall ship by her 268 barrels of ore, which should be insured at $1200 per ton.

"Captain Syme reports that the steamer *Cumberland* passed through the canal some 14 hours before him, but up to the time of writing she has not hove in sight. He further states that the *Mineral Rock* left Copper Harbour two weeks ago, but up to the present there are no tidings from her.

"We have now on the dock 500 barrels of silver ready for shipment. We are now working a very small force of miners, only 23 employed in the mine all told. Six of these are engaged in bailing water and will be dispensed with as soon as I get the new pump which I have ordered.

"I hope my suggestion in regard to advertising for men has been carried out, for as the boats are now running, I shall soon be able to recruit my force."

- Wm. B. Frue, Silver Islet
15th May, 1872

View of the Shangoinah.

The Cemetery.

The Store today.

The big house.

"The mine continues to look well. The *Acadia* unloaded supplies today and I expect she will return on her way below the day after tomorrow, when I shall make shipment of silver which I think should be insured at $1600 per ton. I will also ship 40 barrels of ore which should be insured at $5000 per ton. We have already shipped by the *Manitoba* 176 barrels, and by the *Cumberland* 184 barrels."

- Wm. B. Frue, Silver Islet
20th May, 1872

Aerial view of the mainland.

"I am pleased to state that my faith in the producing capability of Silver Islet was never stronger. The appearance of the mine has improved materially during the past month, the product for October being 150 barrels of rich ore. From present indications, we shall probably pay about $160,000 dividend in January, notwithstanding the enormous cost of permanent improvements which drew heavily on the Treasury - among others, a powerful hoisting and pumping engine for the Islet, which alone will cost $35,000. Many of the stockholders may be disappointed in the result of this year's operations, but as we are now at a sufficient depth to start the proper opening out of the mine, I am in hopes that their expectations will be realized in the coming year. Mining on Burnt Island and the mainland has so far failed in producing any good result."

- Wm. B. Frue, Silver Islet
2nd November, 1872

The Islet, at left, as it looks today.

"We had a narrow escape crossing the lake from Houghton. While rounding the west end of Isle Royale at 7 p.m. on Saturday, October 19, the Captain of the *Silver Spray*, judging he had run a sufficient distance, hauled up on his course for Silver Islet. I was keeping a sharp lookout and perceived low rocks ahead. Immediately the engine was reversed and the wheel put hard astarboard, but we merely escaped Scylla to fall into Charybdis: we had struck on a sunken reef. We remained in this perilous position for an hour and a half, and only after considerable exertion succeeded in disengaging ourselves, arriving safely at Silver Islet at 1 a.m. on Sunday, October 20th. The accident was caused by a local magnetic attraction on the compass deceiving the Captain."

- Wm. B. Frue, Silver Islet
3rd November, 1872

"The weather has been very much against us for the past 3 weeks. The last storm carried away 100 tons of our rich stamp rock. It also picked out of the top crib-work 200 tons of rock, throwing it around the blacksmith's shop. This crib stands 18 feet above the level of the lake, so one may imagine what a southeast gale can accomplish.

"Our work thus far on Burnt Island and the mainland has developed nothing, and owing to the discharge of quite a number of miners from the Islet (for reasons too numerous to mention), I have today withdrawn the entire mining force from the mainland to fill up vacancies on the Islet, which is now fully manned. I have also a strong surface force, a portion of which, if necessary, can be worked in as miners.

"I am unable to make an estimate of the quantity of ore that I shall be able to ship before the close of navigation, not knowing when the last boat will depart."

- Wm. B. Frue, Silver Islet
15th November, 1872

The Store, mid winter.

The Sea Lion

"This morning a Christmas outing was organized: a pleasant trek through the snow to a landmark named the Sea Lion. It is a curious formation of rock jutting out into the lake from the mainland cliff not far from here, and using one's imagination one may say it resembles a large animal with its front feet in the water and its hind feet on the shore."

- CONFESSIONS OF A CORNISH MINER
25th December, 1870

Dance Hall.

"Our last assortment of Cornishmen turned out as troublesome as our Norwegians. Ten of them stole off the day after their arrival. The remainder all struck work after one day underground. As to the reasons offered, they are as many and as opposed to compromise as the quills on a porcupine: poor board, foul air in the mine, and the payment of passage money to this place. Besides these great evils, there were several minor objections, such as the fatness of beef, the sourness of bread, coal damp, loss of appetite, uncivility from the boarding bosses, and so on. As I could not with convenience move Silver Islet to the top of Thunder Cape, I told these men to take their baggage and go ashore. To fill up the vacancy thus caused, I have today sent a man off into the eastern townships of Canada to see what can be done in the way of hunting up miners and labourers, of whom I stand badly in need."

- Wm. B. Frue, Silver Islet
30th July, 1873

Reflections.

"The putting up of the new engine progresses slowly. The boilers are nearly completed, and most of the gauges and fixtures have been attached, but it will be the end of September before steam can be put on. The weather for some time past has been favourable for the sinking of our cribs, and the progress of this work is continuing satisfactorily. I had intended giving a lengthy description of affairs at Silver Islet, but do not at present feel fitted to the task, being slightly under the weather."

- Wm. B. Frue, Silver Islet
30th August, 1873

The Avenue.

"Again, the Equinox. Leaves turn yellow, lake waters darken. Joyous shouts of summer diminish to a few sad farewells. The Sibley mansion stands empty and shuttered, with seagulls on the ridgepole. A chill wind sweeps down the deserted Avenue, where only days ago strolled fashionable young women under bright parasols. The swimming beach, vacated, is presided over by a solitary sandpiper. Silence replaces brouhaha, and the blue skies of August have clouded over. One is sorry for things left unsaid to strangers. The *Chicora, Keewena* and *City of Montreal* will each make only one more trip this year."

- CONFESSIONS OF A CORNISH MINER
21st September, 1873

The beach in autumn.

Net-lifting, early summer.

The Store; the big house.

"At the mine we are able to hold the water in check at the 4th level. The new engine works satisfactorily, although it is quite expensive on fuel. This undoubtedly is partly occasioned by the extra speed at which it is run to keep the mine dry.

"We will ship 50 barrels of silver by the *Manitoba* next week, which will be the last shipment of the season. We will send our first winter mail by that boat, via Marquette or Duluth, whichever port she will winter at."

- Wm. B. Frue, Silver Islet
7th November, 1873

James Booker was the last person ever buried in the Silver Islet Cemetery.

"Judging from the steam rising from the old lake this morning, you would readily come to the conclusion that it was a good place for boiling potatoes in. I might say that since November 1st the weather has been colder and stormier than anything I have seen for the past twenty years.

"Miss Richardson is teaching school, the worthy trustees having failed to obtain a man who would officiate both as teacher and minister."

- Wm. B. Frue, Silver Islet
19th November, 1873

Winter.

"I beg herewith to tender my resignation of the management of the Silver Islet mine, the same to take effect on August 1st, 1875. I leave the position with reluctance, but circumstances over which I have no control dictate and compel the adoption of the present course."

- Wm. B. Frue, Silver Islet
19th July, 1875

Original log house on the Avenue.

Wintering party - circa 1960.

Winter sunrise at Silver Islet.

"In the autumn of 1876, the Silver Islet Mining Company of Lake Superior, the then owner of the Silver Islet Mine and adjacent lands, being unable to discharge its Mortgage Bonds, then past due, was compelled to resort to some alternative, whereby its property could be saved from loss by foreclosure.

"Thereupon the shareholders determined and entered upon a plan of adjustment which would reestablish their interests, though in reduced amounts, would change and extend mortgage indebtedness, provide for continuation of work at the mine, and ultimately discharge all liabilities.

"Under this plan, Messrs. Coe, Langley and Hall, as Trustees for all assenting shareholders, proceeded to organize a new company, styled the Silver Islet Consolidated Mining and Lands Company."

- Edward Learned, Office of the
Silver Islet Consolidated Mining & Lands Co.
4th February, 1879

Jack Drynan - Islet House.

"The unfavourable changes in the Silver Islet vein which occurred in the fall of 1873 continued to the close of 1876. The new levels which had been opened up, the 8th and 9th, proved wholly unproductive. A vast amount of exploratory work by means of the diamond drill also failed to discover any deposits of rich ore. The consequence was, of course, great financial embarrassment, and an almost entire cessation of work during the summer of 1877. Fortunately, in the summer of 1878, rich ore was struck beneath the 4th level, south of the shaft, which yielded 721,632 ounces of silver, a quantity sufficient to rescue the mine temporarily from its embarrassment and provide a working capital of $300,000. The question of the future of Silver Islet is one of very great interest to many, besides those pecuniarily concerned."

- Thomas Macfarlane, Montreal
19th September, 1879

On the beach.

A hundred years ago, there was talk of petitioning thc C.P.R. to build a spur line to Silver Islet. Wouldn't that have been interesting? But nothing ever came of it, and until the road went in, back about 1930, coastal steamers like this handsome two-stacker maintained regular summer schedules.

Rapids Queen, 1892 - 1939

Initial excavation of the communal outhouse at Silver Islet.

Camp Bay, where the fishermen play.

Judge Callaway and friends gathering firewood.